A SUMMARY OF THE PROOFS THAT VACCINATION DOES NOT PREVENT SMALL-POX BUT REALLY INCREASES IT

(1904)

BY

ALFRED RUSSEL WALLACE

British Library Cataloguing-in-Publication Data
A catalogue record for this book is available from the
British Library

Alfred Russel Wallace

Alfred Russel Wallace was born on 8th January 1823 in the village of Llanbadoc, in Monmouthshire, Wales.

At the age of five, Wallace's family moved to Hertford where he later enrolled at Hertford Grammar School. He was educated there until financial difficulties forced his family to withdraw him in 1836. He then boarded with his older brother John before becoming an apprentice to his eldest brother, William, a surveyor. He worked for William for six years until the business declined due to difficult economic conditions.

After a brief period of unemployment, he was hired as a master at the Collegiate School in Leicester to teach drawing, map-making, and surveying. During this time he met the entomologist Henry Bates who inspired Wallace to begin collecting insects. He and bates continued exchanging letters after Wallace left teaching to pursue his surveying career. They corresponded on prominent works of the time such as Charles Darwin's *The Voyage of the Beagle* (1839) and Robert Chamber's *Vestiges of the Natural History of Creation* (1844).

Wallace was inspired by the travelling naturalists of the day and decided to begin his exploration career collecting specimens in the Amazon rainforest. He explored the Rio Negra for four years, making notes on the peoples and languages he encountered as well as the geography, flora, and fauna. On his return voyage his ship, Helen, caught fire and he and the crew were stranded for ten days before being picked up by the Jordeson, a brig travelling from Cuba to London. All of his specimens aboard Helen had been lost.

After a brief stay in England he embarked on a journey to the Malay Archipelago (now Singapore, Malaysia, and Indonesia). During this eight year period he collected more than 126,000 specimens, several thousand of which represented new species to science. While travelling, Wallace refined his thoughts about evolution and in 1858 he outlined his theory of natural selection in an article he sent to Charles Darwin. This was published in the same year along with Darwin's own theory. Wallace eventually published an account of his travels *The Malay Archipelago* in 1869, and it became one of the most popular books of scientific exploration in the 19th century.

Upon his return to England, in 1862, Wallace became a staunch defender of Darwin's landmark work *On the Origin of Species* (1859). He wrote responses to those critical of the theory of natural selection, including 'Remarks on the Rev. S. Haughton's Paper on the Bee's Cell, And on the Origin of Species' (1863) and 'Creation by Law' (1867). The former of these was particularly pleasing to Darwin. Wallace also published important papers such as 'The Origin of Human Races and the Antiquity of Man Deduced from the Theory of 'Natural Selection" (1864) and books, including the much cited *Darwinism* (1889).

Wallace made a huge contribution to the natural sciences and he will continue to be remembered as one of the key figures in the development of evolutionary theory.

Wallace died on 7th November 1913 at the age of 90. He is buried in a small cemetery at Broadstone, Dorset, England.

I. WHY DOCTORS ARE NOT THE BEST JUDGES OF THE RESULTS OF VACCINATION.

(1) In the first place they are interested parties, both pecuniarily and in a much greater degree on account of professional training and prestige. Only three years after vaccination was first introduced, on the recommendation of the heads of the profession, and their expressed conviction that it would give lifelong protection against a terrible disease, Parliament voted Jenner £10,000 in 1802, and £20,000 more in 1807, besides endowing vaccination with £3,000 a year in 1808. From that time doctors as a body were committed to its support; it has been taught for nearly a century as an almost infallible remedy in all our medical schools; and has been for the most part accepted by the public and the legislature as if it were a well-established scientific principle, instead of being as the historian of epidemic diseases--Dr. Creighton--well terms it, a grotesque superstition.

(2) Whether vaccination produces good or bad results can only be determined by its effects on a large scale. We must see whether, during epidemics --at different periods or in different places--small-pox mortality is diminished as compared with that from other diseases in proportion to the total amount of vaccination; and this can be done only by the Statistician, using the best materials--in this country those of our Registrar-Generals.

Two of the greatest medical authorities on vaccination, Sir John Simon and Dr. Guy, F.R.S., have declared this to be necessary. The former, in 1857, in a Parliamentary Report on the *History and Practice of Vaccination*, says: "From individual cases the appeal is to *masses of national experience.*" Dr. Guy, in a celebrated paper published by the Royal Statistical Society, says: "Is vaccination a preventive of small-pox? To this question there is, there can be, no answer except such as is couched in the language of figures." The language of figures is "Statistics"; hence, statisticians, not doctors, are the only good judges of this question. But the last Royal Commission consisted wholly of doctors, lawyers, politicians and country gentlemen, not one trained statistician! Hence, as I have demonstrated in my *Vaccination a Delusion*, they have made the grossest blunders and their Report is absolutely worthless.

II. WHAT IS PROVED BY THE BEST STATISTICAL EVIDENCE AVAILABLE.

(1) The only complete and trustworthy records of mortality and of the causes of death which we possess, are those of the Registrar-Generals for England and Wales, for Scotland and for Ireland; the former from 1838, the two latter from later dates. But for London we have records from a much earlier period--the *Bills of Mortality*, which, though not completely accurate, are yet considered to show the rise and fall of the death-rates from the chief diseases then recognised, with sufficient general accuracy to be very valuable. They are continually appealed to in order to show the enormous improvement in the health of London in the nineteenth as compared with the eighteenth century, and this comparison as regards small-pox is one of the stock arguments of the doctors, and was strongly urged by the Royal Commissioners. It is stated over and over again that, down to the year 1800, small-pox deaths were excessive, but that from the very introduction of vaccination in 1800 they began to decrease, and have been getting less and less ever since. No other disease, it is said, has decreased in the same striking manner.

(2) This being the very foundation of the supposed evidence in favour of vaccination it is necessary to examine it closely, when it will be found to be wholly *worthless*, and to illustrate in a striking manner the complete ignorance of

doctors, and also of the Royal Commissioners, of the very elements of statistical enquiry. This requires some little explanation, though it is really a very simple matter.

In order to be able to study the effect of any alleged cause of improved health of the community, we must compare the death-rates before and after the introduction of the supposed cause of improvement (in this case vaccination), and also compare these with the death-rates from other groups of diseases, and from all causes. These facts are given by the Registrar-General in tables showing the number of deaths each year in each million of the population. Now, small-pox, many fevers, cholera, etc. are what are termed epidemic diseases, which attack large populations at irregular intervals with great severity, while at other times they are far less fatal or more local. Hence the yearly death-rates vary enormously. In 1796 more than 4,000 per million died of small-pox in London, while in the next year there were only about 800, and the following year (1798) over 3,000. Again, in 1870 less than 100 per million died of it, while in 1871 there were about 300, and in 1872 about 2,500. Thus the figures go increasing and decreasing so suddenly and so irregularly, that by taking only a few years at one period, and a few at another, you can show an increase or a decrease according to what you wish to prove. Hence it is often ignorantly said that figures can be made to prove anything. But this is quite untrue. They can often be made to *show* anything, which is quite another matter; but if properly exhibited and compared they lead to one conclusion only; they show *the truth*.

(3) There are a few simple rules for getting at the truth in such statistics as we are now discussing. One is that we must take as long *periods of time* as possible; another is that we must use the *largest* populations available. Two other conditions are almost equally important; we must compare, when possible, equal periods before and after vaccination was introduced; and we must also compare the increase or diminution of *small-pox* with those of *other diseases*, in order to discover whether there is anything exceptional in the decrease of small-pox mortality which requires a peculiar cause to explain it.

But the ever-varying figures in long columns are so confusing to most people, that it is impossible to make anything out of them, and to simplify them, averages have to be taken, showing the deaths every five or every ten years, and in other ways, so as to find out what the figures really mean, and even then, by altering the *periods* or beginning *at different years*, a very different result may often be shown.

(4) By far the best way and that usually adopted by statisticians and mathematicians, is to draw out diagrams by which the whole course of the mortality from each disease or group of diseases can be seen and compared at a glance. From the various elaborate tables given in the Reports of the Royal

11

Commission and from the annual reports of the Registrar-General, I constructed twelve diagrams, each showing the comparative rise or fall of small-pox mortality and other diseases in various places and under different conditions; and all these without exception *demonstrate* either that vaccination has no effect whatever, or that it tends to *increase* rather than decrease *small-pox mortality*. These are all given in my little book "Vaccination a Delusion," which can be had from the National Anti-Vaccination League for 9d. a copy, or 10½d. by post.

(5) As many people do not understand these diagrams I here give a part of one of them in a simplified form in order to render statistical diagrams intelligible to all, and it will serve to show what is the nature of the evidence against vaccination, and also how I prove that the statements made by the doctors and by the Royal Commissioners are not only misleading but absolutely untrue.

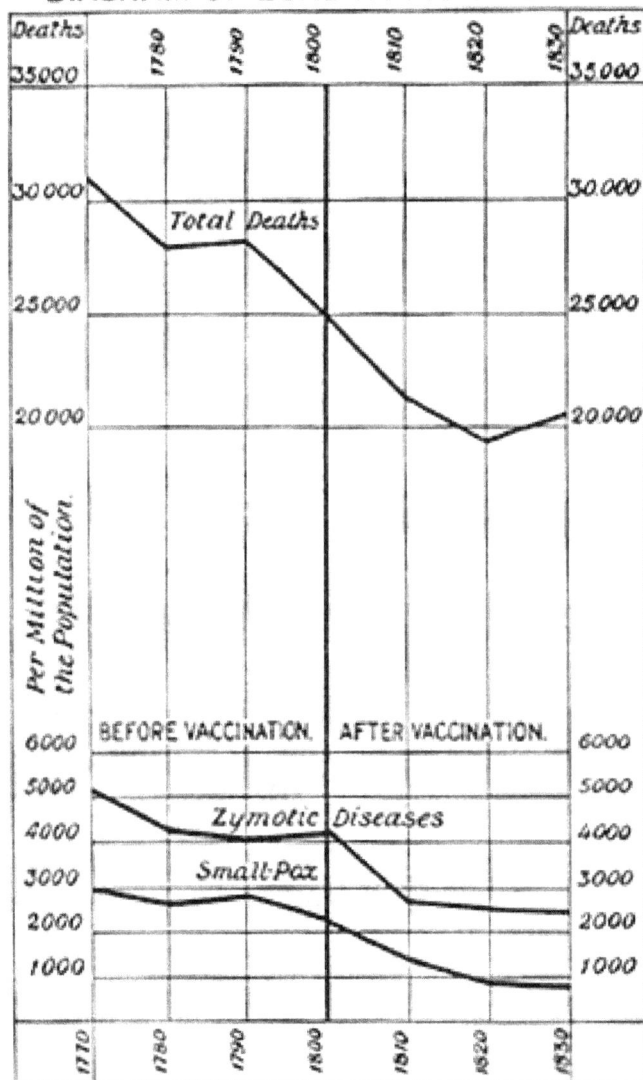

DIAGRAM OF LONDON MORTALITY.

Explanation of the Diagram.

13

(6) The figures on the bottom and top of the diagram show the years, from 1770 to 1830, while those on the right and left show the number of deaths to each million of population. The three wavy lines show the *proportion of deaths* to population during this period of 60 years; the lower line the *small-pox* deaths; that next above it the deaths from the other *zymotic diseases* (fevers, diphtheria, whooping-cough, etc.); while the top line shows deaths from *all diseases*. These last deaths, being so much more numerous, have had to be drawn out on a smaller scale in order to show them on the same page as the others.

(7) This diagram shows us that small-pox decreased during the ten years before vaccination at very nearly the same rate as it did in the ten years after vaccination. The other zymotic diseases decreased even more than small-pox during the ten years after vaccination. General mortality also decreased after 1800 more rapidly than before 1800. Yet the Royal Commissioners declare that there was *nothing* but vaccination to produce the decrease of small-pox, and that there was no improvement in sanitation in the beginning of the nineteenth century, as compared with the latter part of the eighteenth century, to account for the difference.

(8) Now, in an Appendix to my "Vaccination a Delusion," I have given an account of a number of improvements affecting health at this very period which are amply sufficient to produce the results shown by the diagram, and I believe

it is the most compact and most interesting account of these improvements yet given. The chief of them are (1) That many West-end squares and suburbs were built at this very period, and were inhabited chiefly by city people. (2) That the streets were more systematically cleaned and the roads improved. (3) That the water supply was much improved. (4) That potatoes, tea, and coffee came into more general use; while the better roads allowed more fresh meat, vegetables and milk to be used. (5) Cemeteries were formed outside London and many City graveyards were permanently closed. The result of these five groups of improvements was strikingly shown in the decrease of the death-rate in a number of the most fatal diseases (as recorded in a Table by Dr. Farr, reprinted in the Third Report of the Royal Commission) to fully one-half in 1801-10 as compared with 1771-80; an amount of improvement which has never occurred in any similar period during the whole of the 270 years for which we have official statistics. And yet the Royal Commissioners declare that nothing *but vaccination* can explain the corresponding and very similar decrease in small-pox!

(9) As you will now understand the method of exhibiting statistics by means of diagrams, I will proceed to state the other more important conclusions to be drawn from our national statistics of death-rates. Those who wish to study them more fully must obtain the book itself, and examine the diagrams and the full details there given.

III. LONDON DEATH-RATES DURING REGISTRATION. 1838-96.

(1) These tables show us that neither the general mortality nor that from zymotic diseases decreased much till about 1868, but from that date there has been a large and continuous decrease. Small-pox had a sudden increase in 1838, in which year the mortality was greater than it had been for the preceding twenty-five years. Then it decreased slowly till 1870, and this decrease is always ascribed by the doctors to vaccination. But in 1871 there was a great epidemic, when the mortality was greater than at any period during the preceding seventy years of constantly increasing vaccination! Since 1871 small-pox has decreased, but only at about the same rate as the other zymotic diseases. The interesting thing to note here is, that the Main Drainage of London was completed in 1865, and about five years later (the time required for the connection of all the house drainage) the marked diminution in the mortality above-mentioned began to show itself. And if we average the enormous small-pox mortality of 1871 with that of the preceding ten years, we shall find that it will bring the small-pox mortality into almost exact correspondence with that from all other causes, and thus leave nothing to be imputed to vaccination!

(2) In another diagram in my book I show the mortality from the five groups of zymotic diseases taken separately: Fevers, Whooping-cough, Diphtheria and Scarlatina,

Measles, and Small-pox, for the same period of Government Registration. All of these diseases show a nearly similar decrease in the latter half of the period, except measles, which shows hardly any diminution; but there is reason to believe that the cause of this is, that, when vaccinated children after a short illness die of small-pox, measles or chicken-pox are often given as the cause of death.

IV. DEATH-RATES IN ENGLAND AND WALES DURING THE PERIOD OF REGISTRATION.

(1) My third diagram is one of the most instructive and conclusive in my book, because it deals with the whole population of England and Wales and the death-rates from various groups of diseases as in the illustrative diagram. In the first twenty-five years, from 1848 to 1872, there is on the average hardly any decrease either of general mortality, zymotics, or small-pox, since the enormous small-pox mortality of 1871-72 if distributed over the preceding ten years will bring it to correspond closely with the other classes of mortality. But from 1873 to 1895--the last twenty-three years shown--there is a diminution in all three of the diseases to a considerable amount. For the last ten years the diminution in small-pox is the greatest; but this can be proved to be *not* due to vaccination, as I will now explain.

(2) It is only from the year 1872 (after the great epidemic of small-pox) that all vaccinations, private as well as public, have been officially registered, and a table showing their amount has been given in the Final Report of the Royal Commission. From 1872 to 1882 the vaccinations amounted to 95 per cent. of the births; practically *all* were vaccinated if we allow for those that died before they could be operated on or very soon afterwards. But from that date the number of vaccinations steadily decreased, till in 1895

they were only 80 per cent. of the births, a diminution of 15 per cent. in fourteen years. If vaccination were the chief or only preventive of small-pox we ought to have a considerable *increase* of the disease during this period, instead of which it is in this period only that the diminution of small-pox has been *more marked* than that of the other zymotic diseases! Here, then, we have the first distinct proof that it is vaccination which *keeps up the disease*, and that when a larger number of children escape the blood-poisoning lancet small-pox diminishes!

Another and even more conclusive proof is given on page 2 (back of the title-page) by Dr. Ruata, M.D.[1] The whole male population of Italy are revaccinated on entering the army. Under the age of 20, men and women are *alike* as regards vaccination; afterwards men have an *enormous advantage*, if vaccination is of any use. Yet, over 20, many *more* men than women die of small-pox, while under 20 the mortality is equal, again demonstrating that vaccination *increases* small-pox mortality!

19

V. THIRTY YEARS OF RAPIDLY DECREASING VACCINATION IN LEICESTER, AND ITS TEACHINGS.

(1) The great manufacturing town of Leicester, with nearly 200,000 inhabitants, affords the most conclusive proof of the uselessness of vaccination that it is possible to have; and the doctors and government officials carefully avoid dealing with it except to prophecy evils which have never come to pass.

Down to 1872 Leicester was one of the most completely vaccinated towns in the kingdom, the number of vaccinations, owing to alarm after epidemics, several times *exceeding* the number of births. Yet in 1871, at the very height of its good vaccination record, it was attacked by the epidemic with extreme severity, its small-pox deaths during that year being more than 3,500 per million of the population, or about a thousand per million *more* than the mortality in London during the same epidemic. If ever a test experiment existed it is this of Leicester, where an almost completely vaccinated community suffered more than unvaccinated and terribly insanitary London, on the average of the last forty years of the eighteenth century.

But even more conclusive evidence is to come.

(2) That fearful mortality destroyed the faith of Leicester in vaccination. Poor and rich alike, the workers and even the municipal authorities began to refuse vaccination for

their children. This refusal continued till, in 1890, instead of 95 per cent. the vaccinations reached only 5 per cent. of the births! As this ominous decrease of vaccination went on the doctors again and again prophesied against it, that once small-pox was introduced it would run through the town like wildfire and decimate the population. Yet it *has* been introduced again and again, but it has never spread; and from that day to this no town in the kingdom of approximately equal population has had such a very low small-pox mortality as this almost completely unvaccinated and--as the doctors say--*unprotected* population! Surely this completes the demonstration that vaccination, instead of preventing, increases the liability to small-pox, and that the only way to abolish the disease is to do as Leicester did, leave off vaccination altogether and devote our energies to sanitation, and the isolation of such rare cases as do occur.

Yet this wonderfully conclusive test experiment was passed over by the Royal Commissioners in 1894, with a few scattered remarks, which are either absolutely untrue or entirely beside the question. (See "Vaccination a Delusion," 277.)

VI. THE ARMY AND NAVY: A DEMONSTRATION OF THE USELESSNESS OF VACCINATION.

(1) The doctors always claim that, though the effect of vaccination in infancy wears out, yet re-vaccination offers an almost complete protection for the rest of the person's life. In a circular issued in 1884, and up to the time of the Royal Commission widely distributed *with the approval* of the *Local Government Board*, it is stated that: "Soldiers who have been re-vaccinated can live in cities intensely affected by small-pox without themselves suffering to *any appreciable degree* from the disease." I will now show you that this official statement is *absolutely false*.

(2) All soldiers and sailors are re-vaccinated on entering the service, unless they have recently had small-pox. The reports of the Royal Commission give the small-pox deaths in the Army and Navy from 1860 to 1894. The Registrar-General gives the total mortality from disease in the two services for the same period. I have compared these two mortalities by means of a diagram constructed from the tables, and this is what we find. First, throughout the whole period the total mortality from all diseases in the Army is much higher than in the Navy. Clearly, this is the result of the one class living in barracks, largely in towns and cities, the other in the midst of the pure and bracing sea air.

In the second place, there has been, in both services,

throughout the thirty-four years a continuous diminution of mortality, so that it is now only about one-third of what it was thirty-four years ago; and this enormous improvement is stated by the Army and Navy doctors to be due to the much better *sanitation* of ships and barracks, and to the great improvement in the *food* and general *treatment* and *medical* attention in both services.

Thirdly, in both Army and Navy there has been a large decrease in the small-pox mortality throughout the whole period, corresponding closely with that of the general mortality, and certainly due to the same causes--improved sanitation and medical treatment.

Fourthly, in the very same years (1871-2) as the great epidemic in England and on the Continent, there was also a small-pox epidemic both in the Army and the Navy, and taking account of the age of the men and their condition of constant medical supervision, quite as severe as among the general population, who had *not* the alleged *complete protection* of re-vaccination.

Fifthly, this is proved by two comparisons--with Ireland and with Leicester--from tables given in the Reports of the Royal Commission extending from 1864 to 1894. The diagrams formed from these tables show us that Irishmen of about the *same ages* as our soldiers and sailors suffered more during the epidemic of 1872, but for the remainder of the thirty years they had rather less small-pox mortality; while since 1881 they have had *not half the small-pox mortality* of the Army and Navy.

(3) The other comparison is with Leicester, which city, in the period of twenty years (1873-1892), during which they had been growing less and less vaccinated, has had a total of only 16 small-pox deaths per 100,000 of its population, which includes thousands of unvaccinated children and infants; while for the same period the deaths in the Army and Navy amounted to over 70 per 100,000.

And yet we have had the impudently false statement circulated by thousands, under the approval of the Local Government Board, that the re-vaccinated Army and Navy do not, under the worst circumstances, "appreciably suffer!" The Royal Commissioners, on the other hand, shirk the whole matter--make no comparisons with other populations--but state vaguely that "particular classes" who have been "exceptionally" re-vaccinated exhibit "quite exceptional advantages in relation to small-pox,"--a statement which, as regards the only "exceptionally" re-vaccinated large classes of men, is, as their own tables show, the very reverse of the truth, since they suffer much more than the least vaccinated class of about equal population in the whole kingdom.

It is thus absolutely demonstrated that it is the exceptionally *unvaccinated* that possess the exceptional advantages, while the "exceptionally *re-vaccinated*" Army and Navy show quite exceptional *disadvantages*, in a small-pox mortality during the same twenty years, more than *four* times as great as the exceptionally *unvaccinated* town of Leicester!

But the learned men of the Royal Commission never put these two facts side by side, so that the Government and the public might draw their own conclusions from them. So far as their Final Report shows, these gentlemen were ignorant or oblivious of the very existence of these facts, which conclusively prove that Vaccination is not only *worthless* but an *injurious* operation--a Gigantic Medical Imposture!

(4) For the reasons now stated, we call upon voters of all parties to refuse support to every candidate who upholds the legal or other enforcement of vaccination, which, as we have shown, both spreads disease and increases mortality. No government has the right to order healthy infants to be blood-poisoned, under the pretence of protection against a danger that may never arise. The abolition of all laws enforcing or encouraging vaccination is therefore of more immediate and vital importance than any party dogma or any political programme.

Note.--The evidence against vaccination and the misstatements of doctors and officials are fully and clearly set forth in the author's pamphlet--"Vaccination a Delusion," to be obtained from the National Anti-Vaccination League, price 10½d., post free.

VII. HOW TO DEAL WITH MEDICAL PRO-VACCINATORS.

(1) In my "Vaccination a Delusion" I have given examples of the grossest misstatements of doctors and officials from the time of Jenner down to the present day. They are such as often appear to be incredible, but none of them have ever been disproved. Several have been given here; but there is one more which is so universal that it must be briefly referred to. In all Official Reports of small-pox epidemics the fatality of the unvaccinated is always declared to be enormous as compared with the vaccinated. As an example, Dr. Gayton, in a Table published in the Second Report of the Royal Commission, gives the percentage of deaths to cases as follows:--

Vaccinated--7.45 per cent.

Unvaccinated--43.00 per cent.

But all the medical writers on small-pox during the eighteenth century agree in stating that the average death-rate of small-pox patients was then from fourteen to eighteen per cent. At that time, however, the sanitary state of our towns and hospitals was abominable, while the medical treatment of small-pox was so incredibly bad that it is a wonder any survived. Yet the doctors ask us to believe that *now*, with far healthier conditions and with far better treatment and nursing, more than *twice* as many unvaccinated small-pox patients die as died *then*, when *all* were unvaccinated! The thing is absolutely incredible and absurd; and the belief in it is due solely to the fact that doctors register all deaths from small-pox as *"unvaccinated"* when they can possibly find any excuse for doing so. One of them has stated that "the mere assertions of patients or their friends that they were vaccinated counts for nothing." The alleged enormous mortality of the unvaccinated is further shown to be erroneous by the fact that the published Reports of three of the largest small-pox hospitals for London from 1876 to 1879 showed that the average small-pox mortality of all patients was about 18 per cent., or a little higher than during the eighteenth century. This may be explained partly by the fact that many of the milder cases do not go to the hospitals, and partly by the weakening of the constitution due to the blood-poisoning operation of vaccination, which, when conditions are alike, renders the vaccinated less able to resist small-pox than the unvaccinated. It has been well asked: "If about 36 per cent. of unvaccinated patients die of small-pox while only about 18 per cent. died in the eighteenth century who or what kills the other 18 per cent.?" It cannot be the general conditions, since the mortality from all diseases has greatly diminished. There remains only the medical treatment. Do doctors accept this?

(2) Now if any one brings forward doctor's or official's figures as to the enormous value of vaccination, ask them first the above questions. They will deny the facts. Then, in

my book you will find the official authority for these and all the other facts referred to. They will be obliged to say they have never enquired into them, and you may then tell them that they have no right to teach you who *have* enquired into them.

If you have a medical man to deal with, ask him why he does not admit Sir John Simon's statement, that "*the great masses of national experience* can alone prove the value of vaccination." Then show him the diagrams (in my book) which I have here referred to, and ask him to prove that they show "great benefits of vaccination," instead of showing as they do its absolute worthlessness.

(3) As to its terrible dangers, the thousands of lives vaccination has destroyed or ruined as regards health, I have no space to refer to them here, but ample evidence from the Royal Commission Reports is given in my book.

(4) Doctors and Members of Parliament are alike grossly ignorant of the true history of the effects of vaccination. They require to be taught; and nothing is so likely to teach them as to show them the diagrams I have referred to in this short exposition of the subject--those of *London* for thirty years before and after vaccination--of *England and Wales* during the period of official registration--of *Leicester* which has almost abolished small-pox by refusing to be vaccinated for thirty years--and for the *Army and Navy*--which, though thoroughly *re-vaccinated* and therefore (according to the

doctors) as well protected as they possibly can be, yet die of small-pox at least as much as badly vaccinated Ireland, and many times more than unvaccinated Leicester.

A doctor who *has not* studied these most vital statistics has no right to an opinion on this subject.

A candidate for Parliament who *will not* give the necessary time and attention to study them, but is yet ready to vote for penal laws against those who know infinitely more of the question than he does, is utterly unworthy to receive a single vote from any self-respecting constituency.

Note Appearing in the Original Work on :
[1] *communication from Dr. Ruata:* "There is another consideration which has a certain relation with vaccination and small-pox in the Italian Army. Our young men are obliged, by law, to enter the Army at the age of twenty, so that the greatest part of them pay this tribute to the State. The consequence is that, after the age of twenty years, men are by far better vaccinated than women, and after the age of twenty small-pox should kill less men than women. I wished to ascertain if this were true, and here are the figures representing the numbers of deaths from small-pox in men and in women before and after the age of twenty during our great epidemical years, 1887-88-89:--

1887. 1888. 1889. Total.

Deaths: Men / Women

Under Twenty . . . .

5,997 / 5,983   7,349/7,353   5,626/5,631   18,972/18,968

Over Twenty . . . .

2,459 / 1,810   1,990 / 1,418   1,296 / 863   5,745 / 4,091

"All the following years until the last-known (1897) give the same results.

"I had care to send you these facts, which every one can appreciate as he thinks best; and I hope that, for love of truth, you will publish them in the *British Medical Journal.*"

I remain, dear Sir,
Yours most faithfully,

Charles Ruata, M.D.,

Professor of Materia Medica in the University of Perugia, and Professor of Hygiene in the Royal Agricultural College, Universita di Perugia, May 10th, 1899.